DATE DUE

Metro Litho
Oak Forest, IL 60452

93-247

17398-4

537   Robson, Pam
ROB   Electricity

SCIENCE WORKSHOP

ELECTRICITY

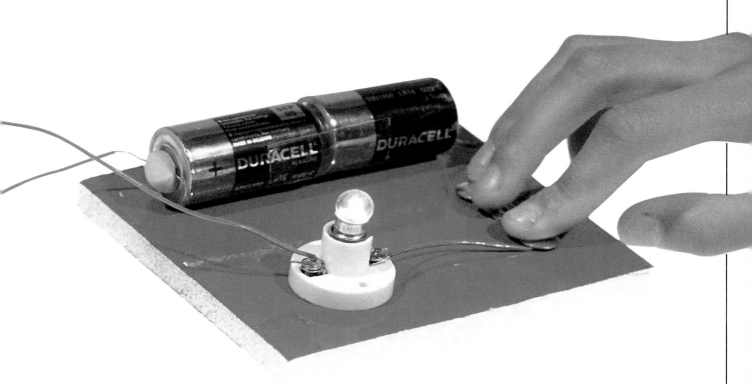

Pam Robson

GLOUCESTER PRESS
NEW YORK CHICAGO LONDON SYDNEY

Design   David West
                    Children's Book Design
Editor   Suzanne Melia
Designer   Steve Woosnam-Savage
Picture Researcher Emma Krikler
Illustrator   Ian Thompson
Consultant   Geoff Leyland

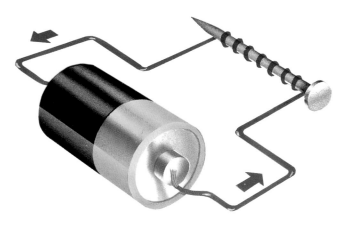

© Aladdin Books 1992

First published in
the United States in 1993 by
Gloucester Press
95 Madison Avenue
New York, NY 10016

Library of Congress
Cataloging-in-Publication Data

   Robson, Pam.
     Electricity / Pam Robson.
       p.   cm. — (Science workshop)
     Includes index.
     Summary: Explores the properties of simple
circuits and inventive ways of putting them to use for
light and power.
     ISBN 0-531-17398-4
     1. Electronic circuits—Juvenile literature. 2.
Electricity—Experiments—Juvenile literature. [1.
Electricity—Experiments. 2. Experiments.] I. Title. II.
Series.
TK7820.R63   1993
537'.078—dc20   92-37099   CIP   AC

CONTENTS

PHOTOCREDITS

All the photographs in this book are by Roger
Vlitos apart from pages: 4 top right: Science
Photo Library; 6 top and 14 top: Eye
Ubiquitous; 10 top: Mary Evans Picture
Library; Frank Spooner Pictures.

INTRODUCTION

Electricity, natural and generated, touches every aspect of our lives. Electricity makes our bodies work as electric currents carry messages along nerves inside us. In some animals, like the electric ray, a strong electric current can be generated to ward off enemies. Electricity can be converted into heat and light, it can magnetize, and it can be transformed into mechanical energy. So many things we do today are made easier by electricity. It enables us to cook, wash, and clean without effort, it powers the machines that entertain us and the trains we travel in. It even allows us to communicate across the world instantaneously. Electrical equipment fills almost every home. Soon we will see a computer that can store X-rays, a compact disc that can record videos, even a videophone. Yet there are still parts of the world where electricity is nonexistent or unreliable. Areas that do not receive main-line electricity may rely on wind, water, or solar power. These sources of energy may be used more extensively in the future. For although electricity is clean to use, it is usually generated by fuels that are not.

Introduction

Why it Works explaining the science ideas

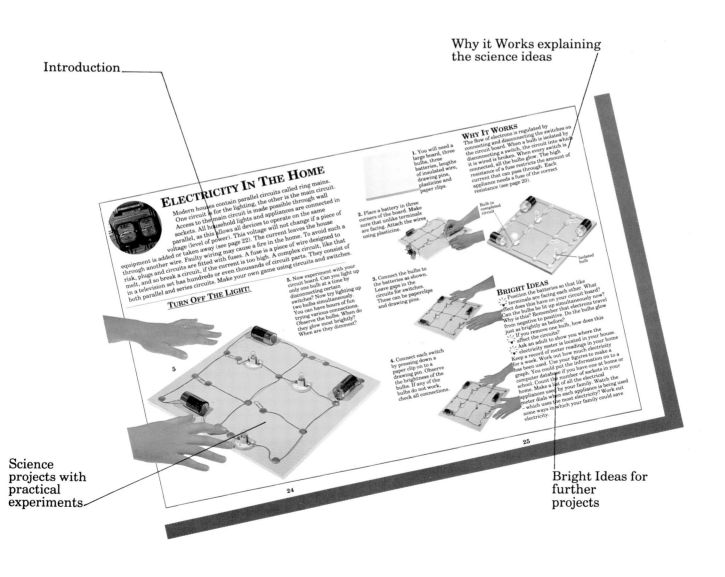

Science projects with practical experiments

Bright Ideas for further projects

THE WORKSHOP

A science workshop is a place to test ideas, perform experiments, and make discoveries. To prove many scientific facts you don't need a lot of fancy equipment. In fact, everything you need for a basic workshop can be found around your home or school. Read through these pages, and then use your imagination to add to your "home laboratory." Make sure that you are aware of the relevant safety rules, and take care of the environment. A science experiment is an activity that involves the use of basic rules to test a hypothesis. A qualitative approach involves observation. A quantitative approach involves measurement. Remember, one of the keys to being a creative scientist is to keep experimenting. This means experimenting with equipment to give you the most accurate results as well as experimenting with ideas. In this way you will build up your workshop as you go along.

MAKING THE MODELS

Before you begin, read through all the steps. Then make a list of the things you need and gather them together. Next, think about the project so that you have a clear idea of what you are about to do. Finally, take your time in putting the pieces together. You will find that your projects work best if you wait while glue or paint dries. If something goes wrong, retrace your steps. And, if you can't fix it, start over again. Every scientist makes mistakes, but the best ones know when to begin again!

GENERAL TIPS

There are at least two parts to every experiment: experimenting with materials and testing a science "fact." If you don't have all the materials, experiment with others instead. For example, if you can't find any polystyrene (a hard plastic) use cardboard or balsa wood instead. Once you've finished experimenting, read your notes thoroughly and think about what happened, evaluating your measurements and observations. See what conclusions you can draw from your results.

SAFETY WARNINGS

Make sure that an adult knows what you are doing at all times. Cutting and bending a coat hanger can be dangerous. Ask an adult to do this for you. In the experiments that use electricity, always use a battery of 1.5 volts. Always make sure your hands are dry. Water and electricity do not mix. Never use main-line electricity. Always be careful with scissors. If you spill any water, wipe it up right away. Slippery surfaces are dangerous. Clean up your workshop when you finish!

EXPERIMENTING

Always conduct a "fair test." This means changing one thing at a time for each stage of an experiment. In this way you can always tell which change caused a different result. As you go along, record what you see and compare it to what you thought would happen. Ask questions such as "why?" "how?" and "what if?" Then test your model and write down the answers.

NATURAL ELECTRICITY

Thousands of years ago, the Greeks noticed that a type of stone called amber attracted light-weight objects, like feathers, after it was rubbed. The Greek word for amber is "elektron." Some materials, such as plastic, do not let electricity pass through, but if they rub against another material, a charge of static electricity can be produced. Static means staying in the same place. We experience static daily – you may hear a crackling sound when you take off your sweater. Sometimes a spark is produced. Rubbing or friction causes static. You can generate your own static and watch the frogs jump.

ACTIVE AMPHIBIANS!

1. Fold a piece of tissue paper a number of times and cut out the shape of a frog. This way you can cut out several frogs at the same time.

1

3

3. Cut a bird shape out of yellow cardboard. Attach it to a Ping Pong ball by threading string through both.

4. Tie the other end of the string to the end of the stick. Make sure that the bird rests on top of the ball.

2. Cut out two lily pad shapes from green cardboard. Cut out some flowers, too. Put the lily pads on the cardboard. Place the frogs on one lily pad.

2

5

5. A short distance away from the tissue paper shapes, rub the ball against the woolen cloth. Do this quite vigorously. This makes the ball negatively charged.

WHY IT WORKS

Most materials are made of atoms that are electrically neutral. However, if atoms gain tiny particles called electrons, they become negatively charged. If they lose electrons, they become positively charged. Like charges will repel, unlike charges will attract. When the Ping Pong ball is rubbed on the woolen cloth, it gains electrons and becomes negatively charged. The tissue paper frogs, which are not charged, jump toward the ball when it is close by. Each time the frogs jump, they are charged in a process called induction.

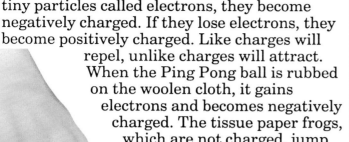

Ping Pong ball

Woolen material

BRIGHT IDEAS

 Rub a balloon on woolen material, then hold it against a door. Now let go of the balloon. Notice what happens. How long does the effect of static electricity last?

Make a water bow. Rub a plastic spoon on wool. Hold the spoon near a stream of water from a tap. The water will bend toward the spoon. See what happens if the water touches the spoon.

6

6. "Fly" the bird above the lily pond. Watch the uncharged frogs leap toward the bird. See if the frogs can leap to the other pad.

THUNDER AND LIGHTNING

Nature's most spectacular display of static electricity is a flash of lightning. Ancient civilizations once believed that thunder and lightning signified the anger of their gods. During a thunderstorm, raindrops and hailstones hurl up and down inside a thunder cloud, producing charges of static electricity. Positive charges move to the top of the cloud, negative charges to the bottom. The earth under the cloud is positively charged, so the negative charges in the clouds are attracted downward. This is why lightning can occasionally strike the earth. The hottest part of lightning can reach a temperature six times hotter than the surface of the sun. You can make sparks like flashes of lightning that are not dangerous at all!

LIGHTNING FLASHES

1. You will need a large plastic bag, a metal tray, modeling clay, and a metal fork or skewer. It is best to do the experiment on a floor with a vinyl covering.

2. Stand the metal tray centrally on top of the large plastic sheet. Put a piece of modeling clay, large enough to use as a "handle," in the center of the tray. Make sure it is firmly secured to the tray.

3. Grip the clay with one hand. Press down firmly and rotate the tray, vigorously, on the plastic sheet. Do this for at least a minute. Using the clay as a handle, lift the tray off the plastic and keep it suspended in the air.

2

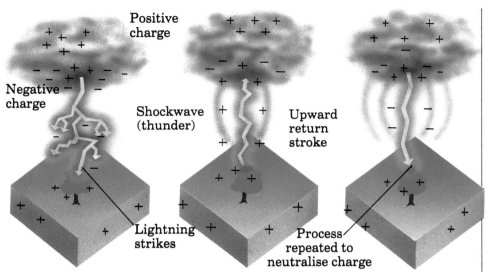

Positive charge

Negative charge

Shockwave (thunder)

Upward return stroke

Lightning strikes

Process repeated to neutralise charge

WHY IT WORKS

Although electricity can pass through the metal tray, it cannot pass through the plastic. As the tray is rubbed on the plastic, it becomes negatively charged. When the positively charged metal fork is brought close to the tray, the negative charges are attracted to the positive. They pass from the tray to the fork as a blue spark. This is how lightning works.

BRIGHT IDEAS

☀ Rub other materials together. Notice which become positively charged and which negatively charged. See if they repel or attract each other. Try rubbing a strip of paper with a woolen cloth. Then hang it over a ruler. The ends of the paper will repel each other because they are similarly charged. Rub a plastic pen with the cloth, then hold the pen between the ends of the paper. It will attract the paper, pulling the ends together because plastic has a strong negative charge. (A moving plastic conveyor belt in a factory can create severe static, and static eliminators have to be used to neutralize the charge.) Unlike charges attract. Try using other materials. Keep a record of your results.

☀ Make a list of materials and the kind of charge created on them – positive or negative.

3

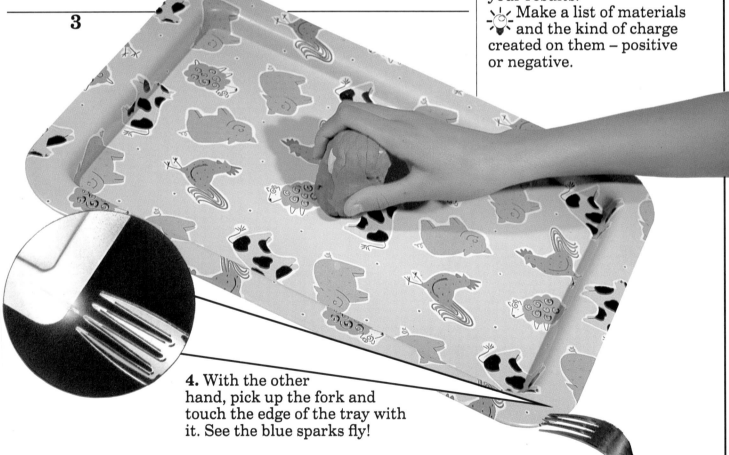

4. With the other hand, pick up the fork and touch the edge of the tray with it. See the blue sparks fly!

A SIMPLE CIRCUIT

Due to the scientific experiments with electricity and magnetism carried out by men like William Faraday, it is now possible to make large amounts of electricity. It can be carried along wires to our homes like water in a pipe, from a power plant. Then it is turned into light, heat, and mechanical energy. Every time you switch on a light you are completing a pathway for the electricity, called a circuit. This allows the current to flow through electrical appliances. It was not until the 1950s that most homes were finally wired up to receive electricity and there are still some areas of the world that have no electricity at all. Our homes are supplied with main-line electricity (see page 24). Do not try to use main-line electricity for the projects in this book – it is dangerous. Using a battery, you can set up a game that needs only a simple, safe electric circuit.

A GAME OF NERVES!

1

2

3. Now ask an adult to open up a wire coat hanger and bend it into bumps and curves. Attach one side of the bulb holder to the bare wire, taking the other end to two batteries.

2. Form a loop with a piece of thin wire, as shown, and connect it to a long length of insulated wire. Put a plastic straw around the join to form a handle.

1. You will need a small 6-volt bulb in a bulb holder. You can buy one of these from a hardware store. Attach two lengths of insulated wire on either side.

3

5. Hide the batteries inside a piece of polystyrene and stand the bent wire on the top. Pass the loop along the wire. If the bulb lights up, try again.

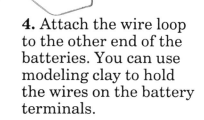

4

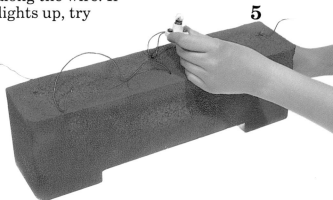

5

4. Attach the wire loop to the other end of the batteries. You can use modeling clay to hold the wires on the battery terminals.

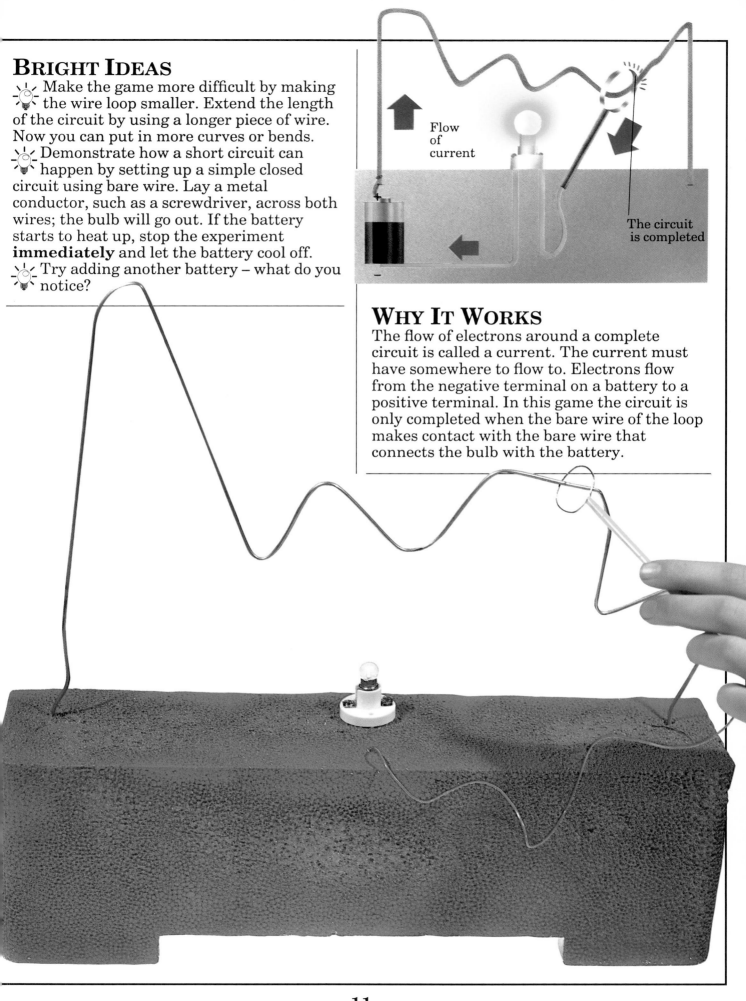

BRIGHT IDEAS

Make the game more difficult by making the wire loop smaller. Extend the length of the circuit by using a longer piece of wire. Now you can put in more curves or bends.

Demonstrate how a short circuit can happen by setting up a simple closed circuit using bare wire. Lay a metal conductor, such as a screwdriver, across both wires; the bulb will go out. If the battery starts to heat up, stop the experiment **immediately** and let the battery cool off.

Try adding another battery – what do you notice?

Flow
of
current

The circuit
is completed

WHY IT WORKS

The flow of electrons around a complete circuit is called a current. The current must have somewhere to flow to. Electrons flow from the negative terminal on a battery to a positive terminal. In this game the circuit is only completed when the bare wire of the loop makes contact with the bare wire that connects the bulb with the battery.

OPEN CIRCUITS

In an open circuit the flow of electricity is controlled by a switch. A switch is the simplest way of controlling the flow of current. When the switch is open, or off, it creates a gap in the circuit. When the switch is closed, or on, the electricity can flow. Telegraphy is the transmission of electric signals. When it was invented in 1838, it allowed people to communicate directly with each other over long distances. In the same year, Samuel Morse introduced his Morse Code – a dot and dash code of short and long electrical signals. These were passed along the wire and decoded at the other end. In 1910, telegraphy was used for the first time to capture a notorious murderer, called Dr. Crippen. The international Morse Code distress signal has always been S.O.S. – three short, three long, three short flashes. Soon ships in distress will transmit a unique identification number via satellite instead.

1. You will need two sheets of polystyrene, modeling clay, two light bulbs and holders, four 1.5v batteries, 6 short lengths of insulated wire, 4 drawing pins, and 2 paper clips.

S.O.S.

2. Place a bulb in each bulb holder and stand them on the polystyrene boards. Connect the two bulb holders and attach wires to the other side of each.

3. Use the drawing pins and paper clips to make a switch for each board. Connect the loose wires to the drawing pins, connecting a paper clip to one of them.

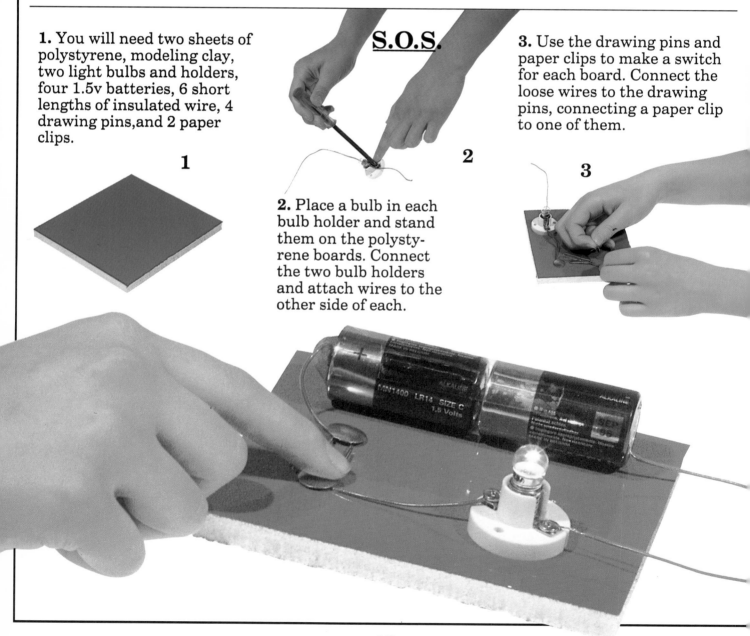

WHY IT WORKS

The paper clips act as switches. Both must be in contact with the drawing pins to allow electrons to flow around the whole circuit, lighting up the bulbs. The sender must raise and lower one paper clip to turn the bulb on and off – the receiver must keep the other down to complete the circuit.

A mechanical switch in the home is slow to work and produces a spark. It joins and separates electrical contacts in the circuit. The spark or arcing produced creates high temperatures. A relay is an electrically controlled switch; it can be operated by various means, but the most common is an electromagnet called a solenoid (see page 28). The solenoid uses an electromagnet to move a metal rod through a short distance. This opens or closes the relay circuit.

(see page 28)

BRIGHT IDEAS

S.O.S. is 3 short, 3 long, 3 short flashes – now try sending a whole message. Find a copy of the Morse Code. Can you "translate" a reply? Can you build a two-way circuit that will work from another room? If you use bulbs, the wiring must be long enough to link the two. See what happens to the light from the bulb if you use longer wire. How can you tell when each word and sentence ends? Try working out a special code, then you can send secret messages.

Make a burglar alarm system with a pressure switch made out of folded aluminum foil. Hide the switch underneath a rug and connect it into an electrical circuit with a buzzer or bulb. Which kind of alarm do you think is most effective – a bulb or buzzer?

How many switches are there in your home? Where are they? What does each switch operate? Carry out a survey.

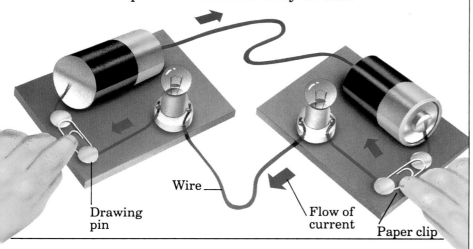

Drawing pin

Wire

Flow of current

Paper clip

4. Rest two batteries end to end on each board and connect them into the circuit using modeling clay. Check to see that the bulbs work by holding down both clips.

5. The bulbs are connected in series. When one bulb lights, both light. A gap in the circuit and the bulbs will not glow.

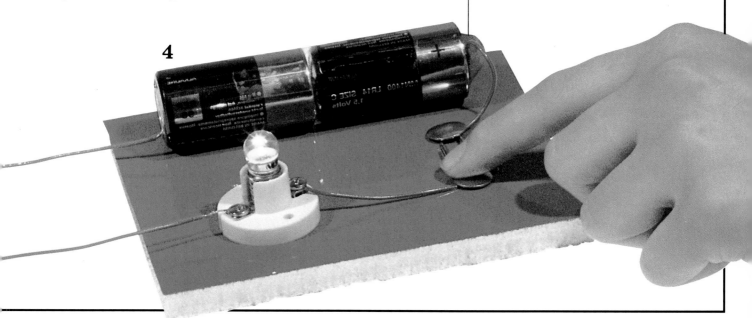

SWITCHES AND ENERGY

Most modern gadgets, such as hair dryers and toasters, switch off automatically. This is not only important to conserve energy, but is a safety precaution too. Heating appliances, such as stoves and irons, have thermostats so that a selected temperature can be maintained – an internal switch turns the heater on or off as required. Because most of our electricity is generated from sources of energy like coal which will run out, we must be aware of the need to save energy whenever possible. The two-way switch is important today, both as a safety device and as a means of conserving energy. If a light can be switched on or off from the top or the bottom of a staircase, not only is it safer at night, but light energy can be saved. Make your own two-way switch and discover how it works.

SWITCH OFF!

1. Position drawing pins at either end of a board so that two plastic lids can just turn, like knobs, within them.

3. Connect one wire to a paper clip inserted through the lid, as shown, taking the other to the batteries.

2. Align four pins as shown and connect them in pairs with two lengths of wire. Make a hole in the top of one plastic knob and insert the two wires connected to a bulb holder.

4. Form a switch at the other lid in the same way, connecting the wire to the batteries. Modeling clay will hold the wires in place.

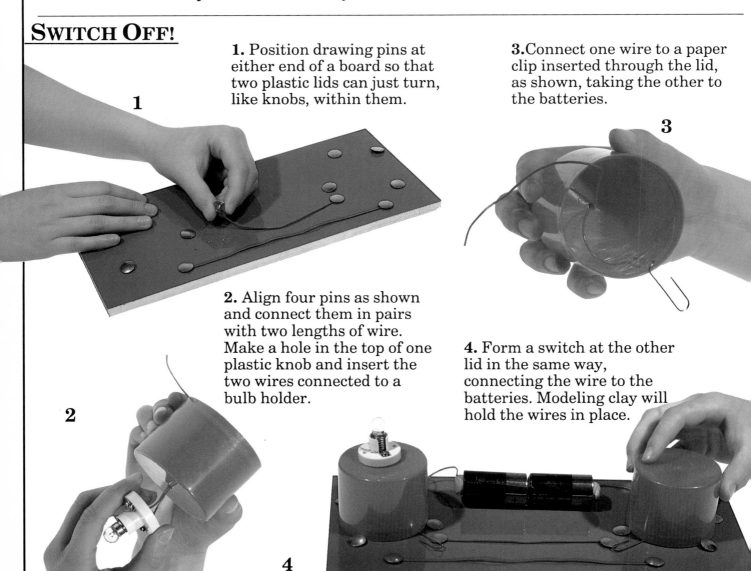

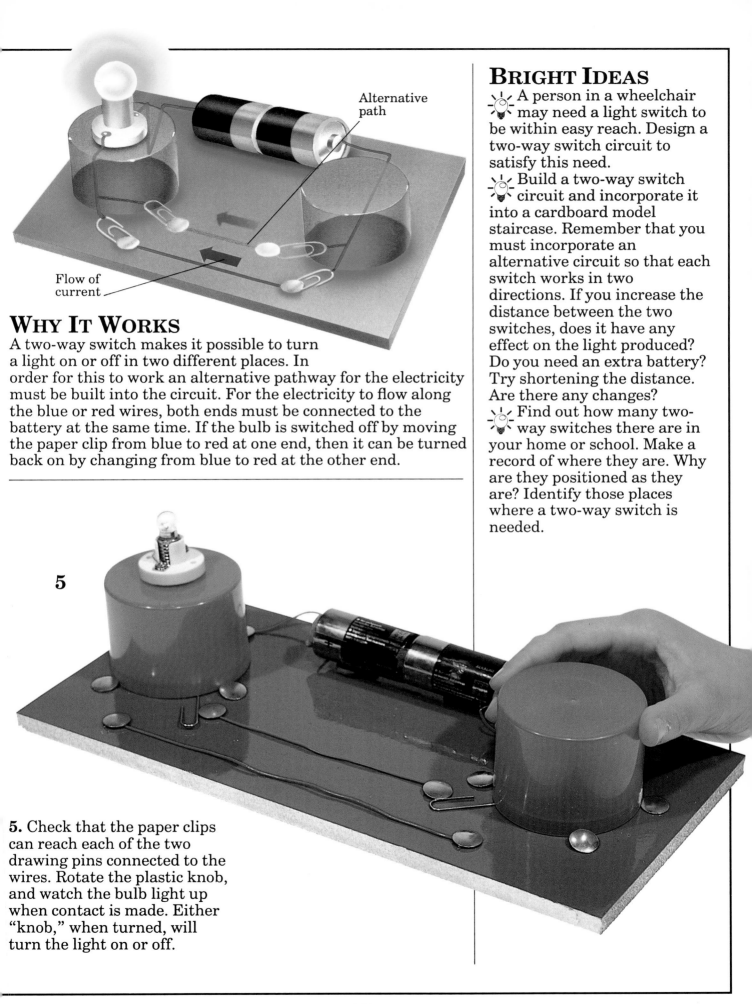

Alternative path

Flow of current

WHY IT WORKS

A two-way switch makes it possible to turn a light on or off in two different places. In order for this to work an alternative pathway for the electricity must be built into the circuit. For the electricity to flow along the blue or red wires, both ends must be connected to the battery at the same time. If the bulb is switched off by moving the paper clip from blue to red at one end, then it can be turned back on by changing from blue to red at the other end.

BRIGHT IDEAS

A person in a wheelchair may need a light switch to be within easy reach. Design a two-way switch circuit to satisfy this need.

Build a two-way switch circuit and incorporate it into a cardboard model staircase. Remember that you must incorporate an alternative circuit so that each switch works in two directions. If you increase the distance between the two switches, does it have any effect on the light produced? Do you need an extra battery? Try shortening the distance. Are there any changes?

Find out how many two-way switches there are in your home or school. Make a record of where they are. Why are they positioned as they are? Identify those places where a two-way switch is needed.

5

5. Check that the paper clips can reach each of the two drawing pins connected to the wires. Rotate the plastic knob, and watch the bulb light up when contact is made. Either "knob," when turned, will turn the light on or off.

BULBS

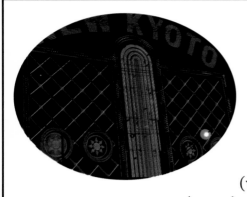

The warning flashes of lighthouses are vital to the safety of ships around the coastline. It was not until the mid-nineteenth century that lighthouses were equipped with electric light bulbs. Two men were responsible for the invention of the incandescent (white-hot) electric light bulb; Thomas Edison, an American, and Joseph Swan, an Englishman. Edison's light bulbs contained a carbon filament within a vacuum. He first produced this on October 21, 1879. By 1913, the tungsten filament (a type of metal) that is still used today had been introduced. Neon lights, like those pictured here, contain a gas. When electricity is passed through the gas, the tube glows. Electronic bulbs have also been developed. These produce only light – not heat – and so save energy.

DANGER AT SEA!

1

3. Use a long cardboard tube for your lighthouse. Cut a piece of polystyrene to fit the end and place the candle through the middle, as shown.

4. Insert the whole thing into the top of your lighthouse, allowing the wires to hang out of the end.

3

4

1. Take a piece of candle and make a hole down the center. Use a paper clip to thread a rubber band through.

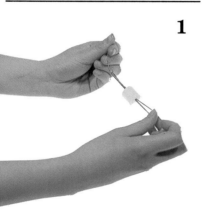

2

5. Fix the polystyrene, candle, and rubber band in place with two toothpicks. Push them right through the cardboard tube, from one side to the other.

6. Line a plastic cup with aluminum foil. Cut out a window to see the bulb. A piece of cardboard with a hole to fit over the bulb will hold it in place.

2. Push the candle into a thread spool and attach the band to the top with tape. Also attach a bulb in a holder to the top, passing the wires down through the candle.

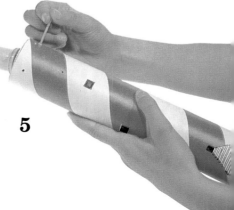

5

16

WHY IT WORKS

As the bulb puts the electrons to work by making it travel through a very long, thin wire called the filament, electric energy is transferred into light energy. Tungsten is a highly resistant metal that can become white hot without melting. Air is removed from the bulb and replaced by the harmless gas, argon. Electrons flow into the bulb when the circuit is complete and cause the wire to glow. Metal at the base of the bulb makes contact with the circuit. Bulbs can become very hot when switched on.

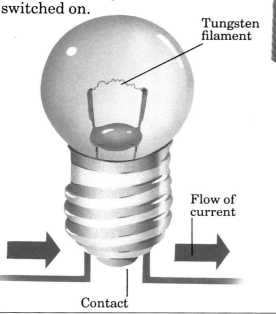

Tungsten filament

Flow of current

Contact

BRIGHT IDEAS

Can you make a different kind of flashing light without switching the current off and on? Adapt the project to make the light revolve and flash in a different way. Try using colored cellophane in the window to make a colored light. Another way to make a flashing light is to use a circle of cardboard, out of which slits like the spokes of a wheel have been cut. Place it in front of the bulb – then revolve the cardboard when the bulb is glowing.

Design and build a traffic light circuit so that the bulbs can be switched on and off in particular sequence. The sequence of change is different in various countries.

Do you know what causes a fluorescent strip light to flicker? The answer has to do with the fact that main-line electricity uses an alternating current (a current that varies all the time).

Design a poster encouraging people to turn lights off and save energy.

7. Connect the wires to a battery and hide it under a papier mâché "rock." Add cotton ball waves. Now twist the thread spool around several times, let go and watch the warning light turn.

7

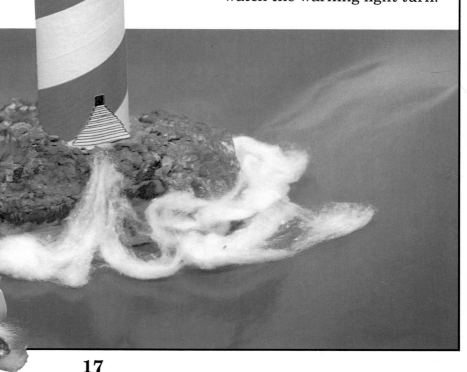

INSULATORS AND CONDUCTORS

Our bodies can conduct electricity, especially when they are wet. Never touch plugs, sockets, or light switches with wet hands. A conductor will allow electricity to pass through it. We use conductors to take electricity to where it is needed. We use insulators to prevent it from reaching places where it could be dangerous. Electricians, like the one pictured here, wear rubber boots to protect themselves from electric shocks. Metal wires conducting electricity are insulated with rubber or plastic to make them safe. Conduct your way through a maze, using insulators and conductors as your guides.

AMAZING!

3. Before you stick down the final part of your "pathway," make a hole near the edge of the board and insert the end of some insulated wire.

1

1. Take a piece of thick board and cut out a piece of aluminum foil of the same size. Cover this with adhesivebacked plastic.

2. Design your maze on the board, and cut out strips of plastic-covered foil to fit the paths. Stick them down making sure that your "pathway" is the last to go on.

2

4. Attach this wire to one terminal of a battery. To the other terminal, attach more wire with a bulb holder in the middle. Connect the other end to a nail.

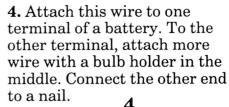

4

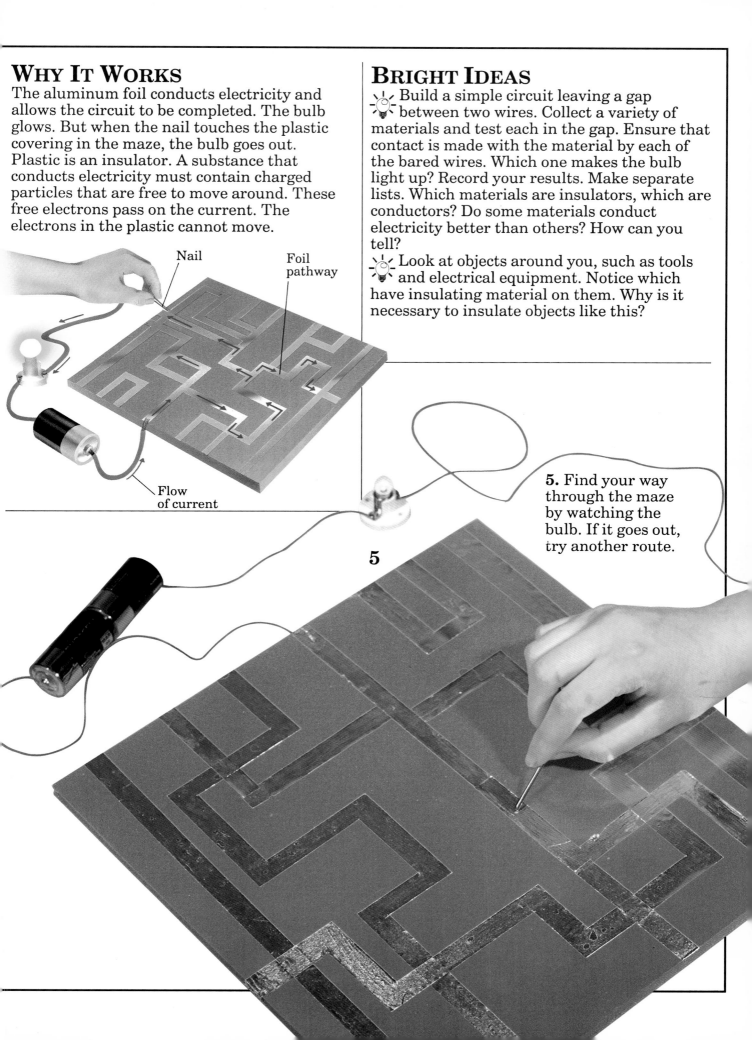

WHY IT WORKS

The aluminum foil conducts electricity and allows the circuit to be completed. The bulb glows. But when the nail touches the plastic covering in the maze, the bulb goes out. Plastic is an insulator. A substance that conducts electricity must contain charged particles that are free to move around. These free electrons pass on the current. The electrons in the plastic cannot move.

Nail

Foil pathway

Flow of current

BRIGHT IDEAS

Build a simple circuit leaving a gap between two wires. Collect a variety of materials and test each in the gap. Ensure that contact is made with the material by each of the bared wires. Which one makes the bulb light up? Record your results. Make separate lists. Which materials are insulators, which are conductors? Do some materials conduct electricity better than others? How can you tell?

Look at objects around you, such as tools and electrical equipment. Notice which have insulating material on them. Why is it necessary to insulate objects like this?

5. Find your way through the maze by watching the bulb. If it goes out, try another route.

5

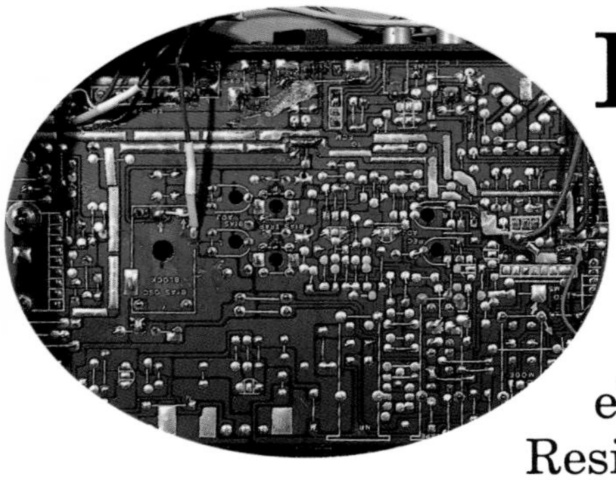

ELECTRICAL RESISTANCE

James Watt, inventor of the steam engine, gave his name to the measurement of electrical power, the watt. Electrical resistance is what makes the filament (long, thin piece of wire) in a bulb glow and the element in an electric heater become hot. Resistance is measured in ohms, after the German scientist, Georg Simon Ohm. Resistors are used in circuits, like the ones pictured here. They are coils of wire, or poor conductors, built into the circuit to reduce the current. A variable resistor, or rheostat, is used to control the speed of a toy car and the volume of a radio or television. By building your own resistor, you can make a night-light with a dimmer switch.

DIM THE LIGHT

1. Set up a circuit using a bulb in a holder, insulated wire, a board, and safety pins. Ask an adult to cut the lead out of a pencil, and cut out two pieces of thick cardboard to rest it on.

1

2. Attach one end of the wire to the batteries and the other end to the pencil lead.

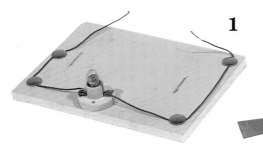

2

3. Rest the lead on the cardboard, and place the batteries with opposite terminals together. Join a piece of wire to a paper clip, and attach it to the other end of the batteries.

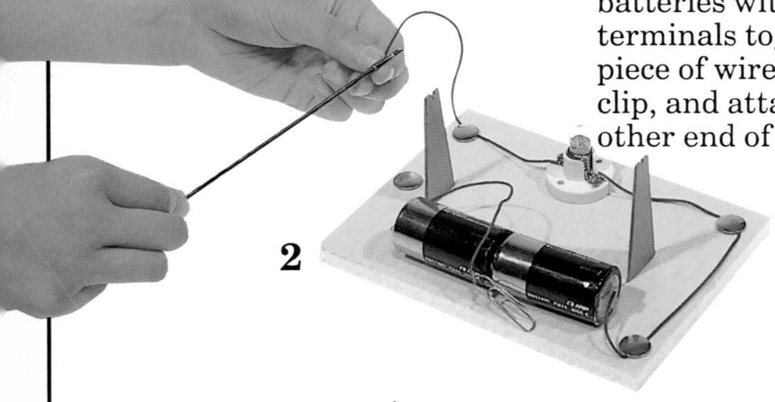

3

4

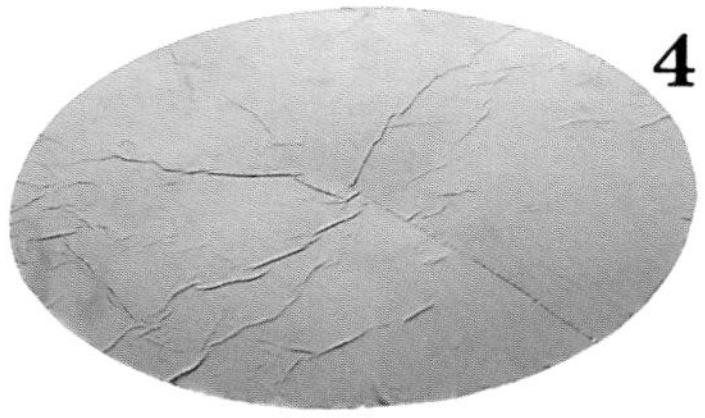

4. To reflect the light, make a shade of cardboard and aluminum foil. Cut a slit in the flat disk and glue it into a cone shape.

5

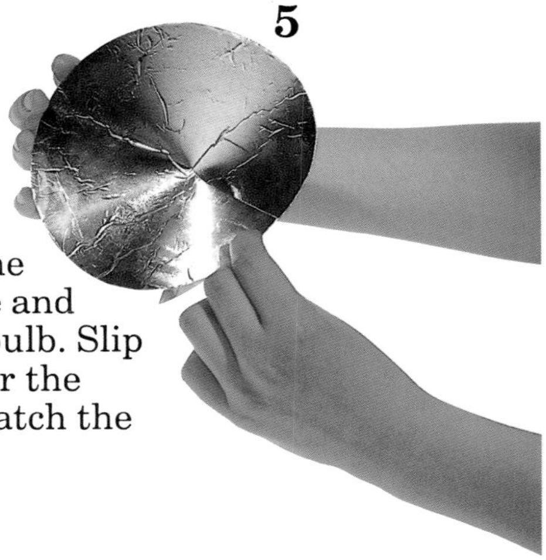

5. Cut a hole in the center of the cone and place it over the bulb. Slip the paper clip over the pencil lead and watch the bulb light up.

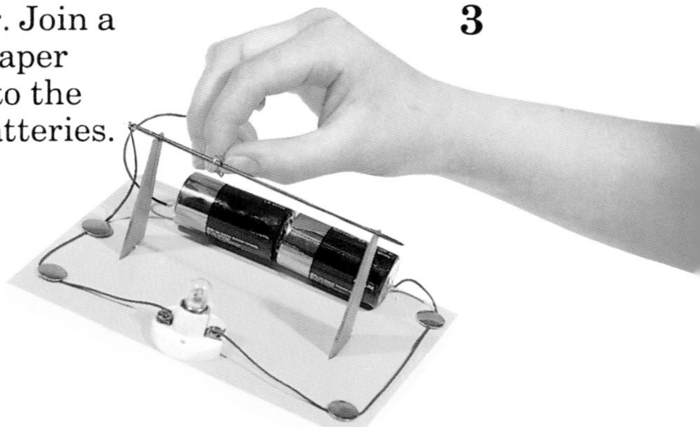

WHY IT WORKS

A pencil lead is made of carbon, which conducts electricity. All conductors have some resistance which becomes higher the farther the electricity has to travel. As the paper clip moves toward the battery, the electricity doesn't have to travel as far. The bulb therefore becomes brighter. As it is moved away from the battery, the light dims.

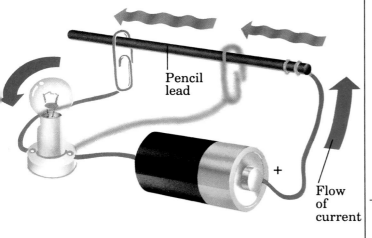

Pencil lead

+

Flow of current

BRIGHT IDEAS

☀ Repeat the project attaching the paper clip to the wire from the bulb. Attach the pencil lead to the wire from the battery. When is the bulb brightest? Which way must you move the paper clip to dim the light now? Which works best?

☀ Build a model theater set with a circuit of floor lights. Use colored paper to create colored lights. By building a variable resistor into the circuit you can dim or brighten the stage lights.

☀ You can make another kind of dimmer by immersing a length of aluminum foil in salt water while it is connected to a circuit. A second piece of foil, connected to the other end of the circuit, is at the bottom of the container. Watch what happens when you move the top piece of foil up and down in the water.

6. Move the paper clip up and down the pencil lead. The bulb should get brighter or dimmer.

6

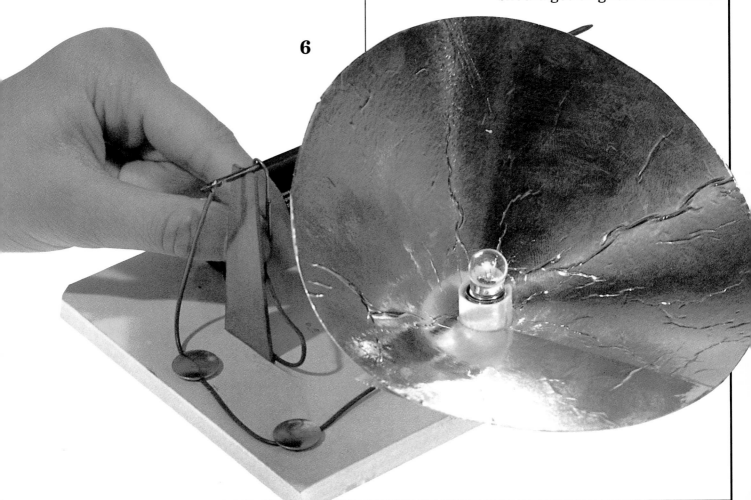

SERIES AND PARALLELS

As long ago as 1810, many larger cities had street lighting. An electric current was made to jump between two carbon rods – this was called electric arc lighting. First introduced by Sir Humphry Davy, these lamps were connected in series. This meant that all the lamps were connected as a part of one large circuit. It also meant that if one lamp went out, and the circuit was broken, they all went out. This often happens with Christmas tree lights, although they can be arranged in parallel circuits to avoid this problem. It was Thomas Edison who recognized the need to use parallel circuits for street lighting. Each bulb in a parallel circuit has a circuit of its own. If one bulb fails, the others will continue to glow; the current is divided equally between them.

LOTS OF LIGHTS

1

1. You will need two large boards, drawing pins, insulated wire, bulbs, bulb holders, and batteries. The drawing pins can act as contacts where your wires join.

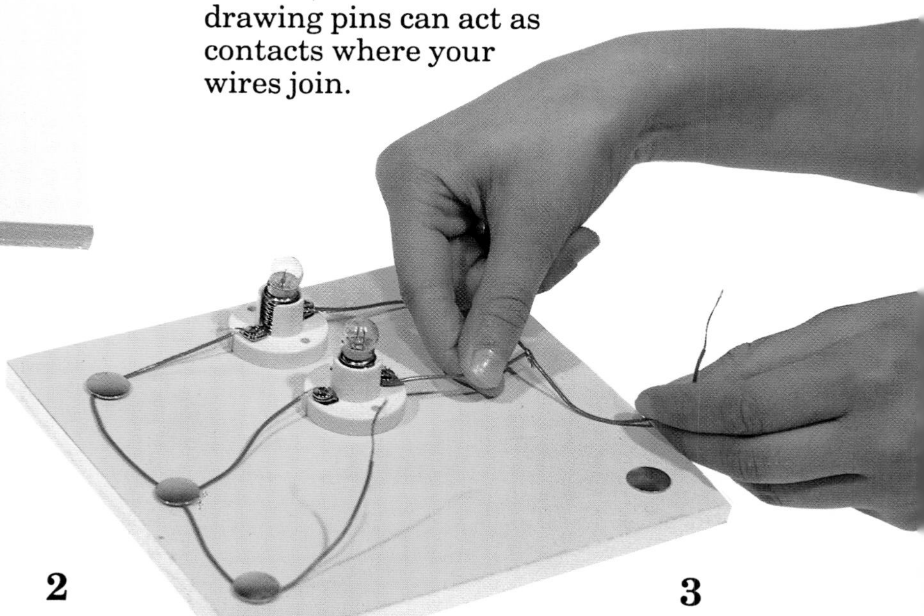

2. Set up your parallel circuit. If one bulb fails the other will remain lit because the circuits are separate. Observe how brightly the bulbs glow.

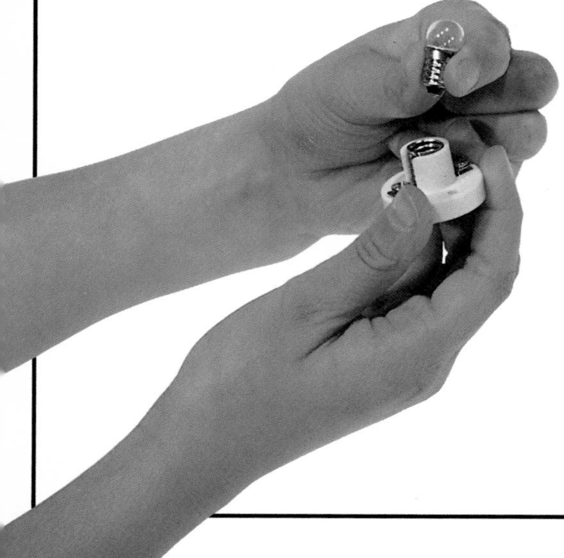

2

3

3. Replace one of the bulbs in the parallel circuit with another battery. Does the light from the bulb change? Now wire up a series circuit like the red one shown here. Include one bulb and two batteries in this circuit. Stop **immediately** if the batteries heat up.

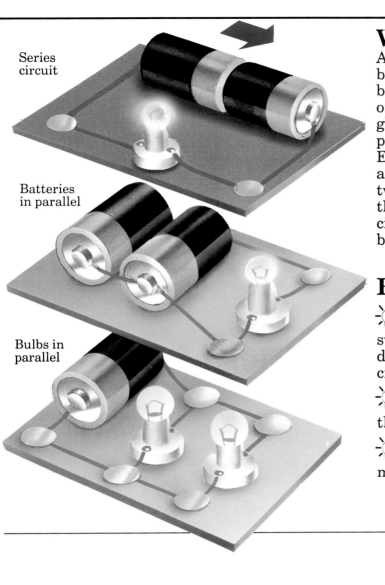

Series circuit

Batteries in parallel

Bulbs in parallel

WHY IT WORKS

A series circuit uses one path to connect the bulb and battery. If two batteries are used, the bulb glows twice as brightly as it would with one. Two bulbs in a series circuit would not glow as brightly as one. A parallel circuit provides more than one path for the current. Each bulb receives the same voltage even if another battery or bulb is added or removed. If two batteries are used in a parallel circuit, their power does not combine as in the series circuit. The bulb receives the voltage of one battery, but glows for double the time

BRIGHT IDEAS

Add another bulb to the series circuit. What do you notice when the current is switched on? Now add another one. What difference does this make? Draw a series circuit diagram.

Wire another bulb into the parallel circuit. What do you notice about the glow from the bulbs? Draw a parallel circuit diagram.

For how long do the bulbs in each kind of circuit stay lit? Which type of circuit is most wasteful of energy?

4

4. Observe this bulb. Does it shine as brightly as the bulb in the yellow parallel circuit? Try removing one battery. Which bulb is shining the brightest now?

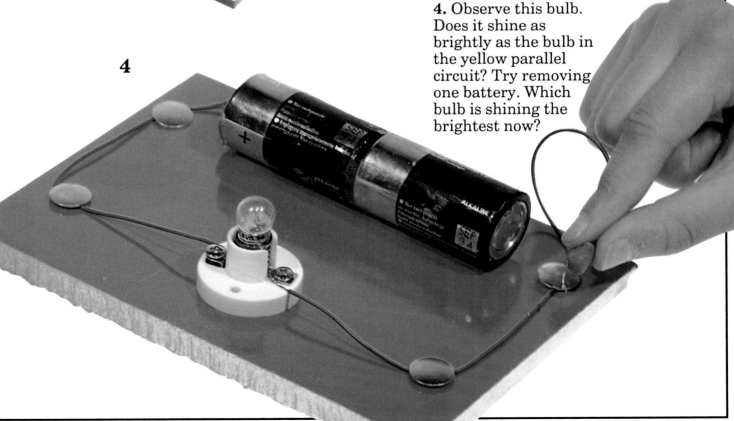

ELECTRICITY IN THE HOME

Modern houses contain many electric circuits. Some circuits are for lighting and others power appliances or heaters. Access to the main circuit is made possible through wall sockets. All household lights and appliances are connected in parallel, as this allows all devices to operate on the same voltage. This voltage will not change if a piece of equipment is added or taken away (see page 22). The current leaves the house through another wire. Faulty wiring may cause a fire in the home. To avoid such a risk, plugs and circuits are fitted with fuses or circuit breakers. A fuse is a piece of wire designed to melt, and so break a circuit, if the current is too high. A complex circuit, like that in a television set, has hundreds or even thousands of circuit parts. They consist of both parallel and series circuits. Make your own game using circuits and switches.

TURN OFF THE LIGHT!

5. Now experiment with your circuit board. Can you light up only one bulb at a time by disconnecting certain switches? Now try lighting up two bulbs simultaneously. You can have hours of fun trying various connections. Observe the bulbs. When do they glow most brightly? When are they dimmest?

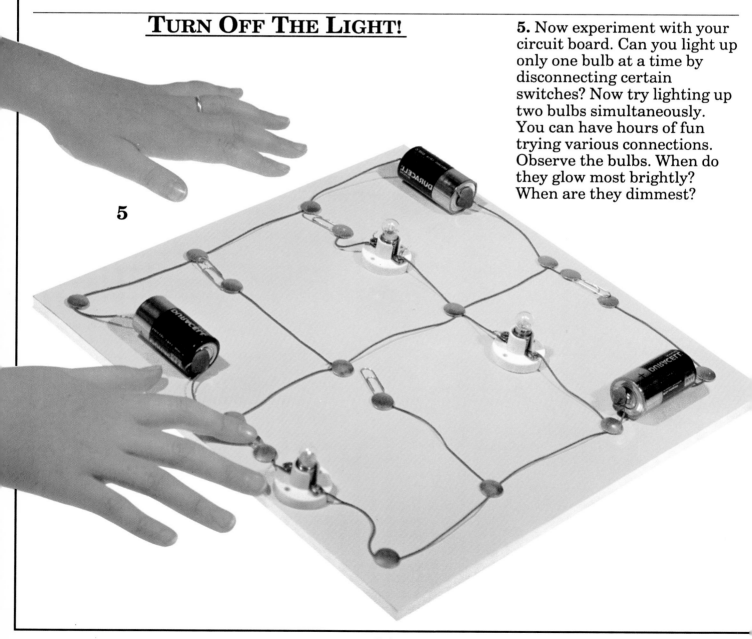

5

1. You will need a large board, three bulbs, three batteries, lengths of insulated wire, drawing pins, modeling clay, and paper clips.

2. Place a battery in three corners of the board. Make sure that unlike terminals are facing. Attach the wires using modeling clay.

3. Connect the bulbs to the batteries as shown. Leave gaps in the circuits for switches. These can be paper clips and drawing pins.

4. Connect each switch by pressing down a paper clip on to a drawing pin. Observe the brightness of the bulbs. If any of the bulbs do not work, check all connections.

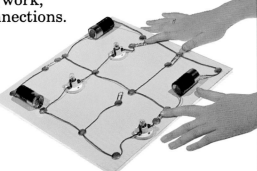

WHY IT WORKS
The flow of electrons is regulated by connecting and disconnecting the switches on the circuit board. When a bulb is isolated by disconnecting a switch, the circuit into which it is wired is broken. When every switch is connected, all the bulbs glow. The high resistance of a fuse restricts the amount of current that can pass through. Each appliance needs a fuse of the correct resistance (see page 20).

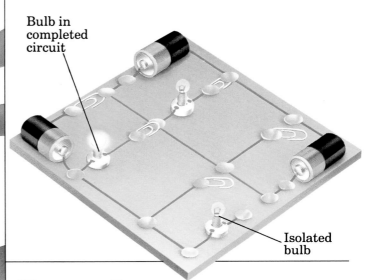

Bulb in completed circuit

Isolated bulb

BRIGHT IDEAS
Position the batteries so that like terminals are facing each other. What effect does this have on your circuit board? Can the bulbs be lit up simultaneously now? Why is this? Remember that electrons travel from negative to positive. Do the bulbs glow just as brightly as before?

If you remove one bulb, how does this affect the circuits?

Ask an adult to show you where the electricity meter is located in your house. Keep a record of meter readings in your home for a week. Figure out how much electricity has been used. Use your figures to make a graph. You could put the information on a computer database if you have one at home or school. Count the number of sockets in your home. Make a list of all the electrical appliances used by your family. Watch the meter dials when each appliance is being used; which uses the most electricity? Figure out some ways in which your family could save electricity.

ELECTRICITY AND MAGNETISM

Hans Oersted, the nineteenth century Danish scientist, first proved the relationship between electricity and magnetism when he noticed that a magnet held near to a compass caused it to turn. When this experiment was repeated, replacing the magnet with a current of electricity, he observed the same effect. This was the beginning of electromagnetism. After Oersted's experiments, it was soon realized that magnets could be made by passing an electric current through coils of wire. The magnetic field (the region around the wire where the force of magnetism is felt) could be switched on and off with the electricity. When a doorbell is pressed, an electromagnet attracts a clapper to strike the bell. Use electromagnetism to hold the clown's nose in place.

RED NOSE DAY

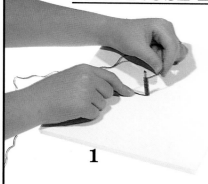

1

1. Take a piece of thick board and push a nail through the center. Now wind a piece of wire around the nail at least 20 times, leaving two ends of the same length.

2. Cut two triangular pieces of polystyrene to support the board in a sloping position.

2

3. Attach the triangles, as shown, and pierce a small hole in the side of one of them for a paper clip to fit through.

3

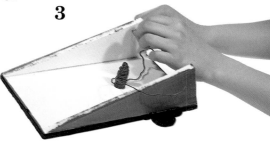

4. On a second sheet of polystyrene draw a clown's face to fit on top of the shape you have made. Do not draw a nose on the face. Color the face, then cut out your clown.

4

5. Affix a drawing pin to the side of a Ping-Pong ball, colored red. This will be your clown's nose. It will not fit in place yet.

5

6. Position the batteries inside the shape as shown. Make sure unlike terminals are touching. Now connect a wire from the nail to one end of the batteries using modeling clay. Connect the other to the paper clip. Stop **immediately** if the batteries heat up.

6

BRIGHT IDEAS

☼ Reproduce Oersted's experiment. Magnetize a needle by rubbing it in one direction on a strong magnet, and rest it on a piece of folded cardboard that is balancing on a stick. Place it in a jar; this will act as a compass. Now set up a simple circuit, allowing the wire to run above the magnetized needle. Observe the effect on the needle when the current flows. Wind more lengths of wire around your compass. What difference does this make?

☼ Find out which appliances, such as a telephone, contain electromagnets.

☼ Can you design a burglar alarm that works because of the effect of an electromagnet?

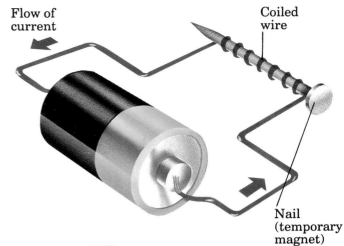

Flow of current

Coiled wire

Nail (temporary magnet)

7. Push the paper clip through the hole until it touches the adjacent battery terminal. Leave half of the paper clip protruding through the hole. Position the red nose on the clown's face.

WHY IT WORKS

When the current is switched on, the nail becomes a temporary magnet. The clown's nose stays in place, held in the magnetic field created by the electricity. When the electric current is turned off, the nose falls off. The nail loses its magnetic properties.

7

8. While the paper clip is pushed inward, the nose will stick to the clown's face. Now pull the paper clip outward. The clown's nose will roll off his face.

8

ELECTROMAGNETISM

The English physicist, Michael Faraday, discovered that electrical energy could be turned into mechanical energy (movement) by using magnetism. He used a cylindrical coil of wire, called a solenoid, to create a simple electric motor. He went on to discover that mechanical energy can be converted into electrical energy – the reverse of the principle of the electric motor. His work led to the development of the dynamo, or generator. You can make a powerful electromagnet by passing electricity through a coil of wire wrapped many times around a nail (see page 26). Electromagnets are found in many everyday machines and gadgets. An MRI scanner (Magnetic Resonance Imaging), like the one pictured here, contains many ring-shaped electromagnets. With a solenoid and a current of electricity, you can close the cage.

CAGED!

1. Take a piece of polystyrene and edge it with cardboard. Stick plastic straws upright around three sides as the bars of the cage.

1

2. Cut out another piece of polystrene of the same size for the roof of the cage. Attach a piece of plastic straw to the side above the door. Wind a piece of wire around a nail 50 times leaving two ends. Affix the nail to the roof, as shown.

2

3. Insert a needle into the straw so that it almost touches the nail. Cut out a rectangle of plastic for the door. Make a hole at the bottom of the door for the needle to fit through.

3

4

4. Stick a piece of cardboard across the door to help hold it open, and make sure the end of the needle just pokes through the hole. Now attach one of the wires to one terminal on the battery. Leave the other free. Make sure it will reach the other terminal. Put the animal into the cage.

WHY IT WORKS

When the current is switched on, the nail becomes magnetized as the current flows through the wire. The needle in the door of the cage is attracted to the electromagnet. As the needle is pulled toward the nail, the door closes to trap the tiger.

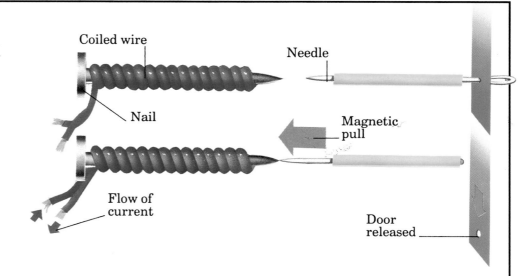

Coiled wire

Needle

Nail

Magnetic pull

Flow of current

Door released

BRIGHT IDEAS

Wind more turns of wire onto the electromagnet. The magnetic effect will increase. What happens if you use a more powerful battery? (Do not let it get too hot.)

Make another electro-magnet using a shorter nail. This will also make the magnetic pull stronger.

Make an electromagnetic pickup by winding wire around a nail. What objects can you pick up? What happens when the current is turned off?

Use an electromagnet to make a carousel spin. Attach paper clips around the edge of a circular cardboard lid to be the roof. Make sure it is free to spin, and place an electromagnet close to the paper clips. The carousel should turn as you switch the current on and off quickly.

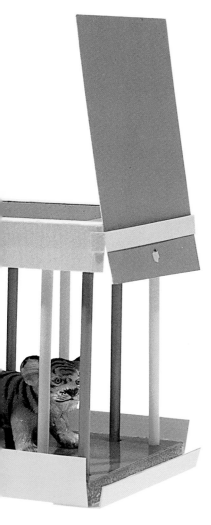

5. Now pick up the free wire. Allow the free wire to come into contact with the unconnected battery terminal. The needle should be pulled back toward the nail. The door will fall down, trapping the animal in its cage.

ELECTROLYSIS

Electrolysis is a process in which an electric current is passed through a liquid, causing a chemical reaction to take place. The liquid used is called the electrolyte. The wires or plates where the current enters or leaves the liquid are called electrodes. The electrolysis of metallic solutions is useful in putting metal coatings on objects. If you have a look at some car bumpers, you will notice that they may have a nice, smooth, metallic appearance. This is because they are coated with a metal called nickel, in a process called electroplating. This helps to stop the metal underneath from rusting. The same method is used to coat cutlery with silver. This is called silverplating. Michael Faraday discovered the first law of electrolysis. The process is also used to purify metals like aluminum.

COPPER PLATING

1. For this project you will need a glass jar, a copper coin, a paper clip, two batteries, insulated wire, and water. Pour the water into the jar. Place the batteries together with unlike terminals adjacent. Connect wires to the terminals. Attach the copper to the wire from the positive terminal of the battery. The paper clip must be attached to the wire from the negative terminal. Use modeling clay. Do not allow the metal objects to touch in the solution. You could even tape each wire to the side of the jar so that they are suspended.

1

2. Observe closely what happens. Can you see bubbles? Leave them for a few minutes, then remove. Observe any color changes. Replace them for a while. Are there any further changes?

WHY IT WORKS

The copper coin is connected to the positive terminal of the battery – the current enters here. The other, the paper clip, is joined to the negative terminal – the current leaves here. As the current flows through the water from the positive electrode (anode) to the negative electrode (cathode), the copper is carried from the coin to the clip.

Movement of copper

BRIGHT IDEAS

 Repeat the project using salt dissolved in vinegar instead of the water. What difference do you notice – if any? What do you observe about the appearance of the paper clip? Maybe your school has scales that can weigh very small objects? If the coin and the paper clip are weighed before immersion in the liquid and their weight recorded, you can check whether electroplating has really taken place. After carrying out the project weigh them both again. Now replace the battery with a more powerful one, or add a second battery into a parallel circuit, to increase the "push" of the current passing through the liquid. (Remember to stop your experiments if the batteries heat up.) Weigh the coin and paper clip a second time. If the weight of the paper clip has increased further, then you have proved the first law of electrolysis – the size of the charge passed through the liquid determines the amount of copper freed.

2

Scientific Terms

ALTERNATING CURRENT
An electric current that reverses its direction around a circuit at regular intervals.

DIRECT CURRENT
An electric current that always flows in the same direction – like that produced in a battery.

ELECTRIC CURRENT
A continuous flow of electrons through a conductor – measured in amperes (amps).

ELECTRIC POWER
The rate of transfer of electrical energy into another form of energy, such as light or heat – measured in watts.

ELECTRICAL RESISTANCE
The degree to which materials obstruct the flow of an electric current – measured in ohms.

ELECTROLYTE
A liquid in which a chemical reaction, electrolysis, takes place when an electric current is passed through it.

ELECTROMAGNETISM
The relationship between electricty and magnetism. Either can be produced from the other.

GROUND WIRE
A wire used as a safety precaution, connecting a piece of domestic apparatus to the ground. If the apparatus malfunctions, the "live" object is "earthed."

INCANDESCENT LIGHT
Light that results when a solid, like tungsten, is heated.

MAIN-LINE ELECTRICITY
The electricity produced by electromagnetic induction. Produced at power plants, it can reach almost every home along power lines.

SHORT CIRCUIT
A weak connection, possibly due to a faulty electrical connection, causing an electric current to take a path of low resistance.

VARIABLE RESISTOR (RHEOSTAT)
A device to control the amount of electricity flowing through a circuit.

Index